Mathematical
Puzzles
and
Pastimes

Edited by Philip Haber
Illustrated by Stanley Wyatt

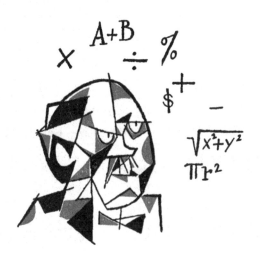

DOVER PUBLICATIONS
Garden City, New York

Bibliographical Note

This Dover edition, first published in 2023, is an unabridged republication of the work originally published by the Peter Pauper Press, Mount Vernon, New York, in 1957. The artwork has been reproduced in black and white for this edition.

International Standard Book Number

ISBN-13: 978-0-486-85139-6
ISBN-10: 0-486-85139-7

Manufactured in the United States of America
85139701   2023
www.doverpublications.com

PREFACE

THE author herewith presents for the enjoyment of the philomath and all other interested persons a collection of recreational mathematics, which can be solved primarily by clear thinking, arithmetic or algebra.

Altogether this book represents over twenty years' constant research in collecting those problems that seemed most interesting, intriguing and universally appealing. Problems were selected from the author's private collection and original works.

A total of 113 problems with answers is given including Four Famous Problems. Some of the problems have solutions, while others are without, leaving it to the reader to rack his brains mulling over the whys and wherefores.

It is sincerely hoped that the reader will thoroughly enjoy himself puzzling and striving for that eternally elusive answer seemingly at hand, yet so far away.

PHILIP HABER

Mathematical Puzzles and Pastimes

1. This man is all mixed up. Someone told him the length of his head is exceeded by the length of his legs by 20½ inches and of his body by 23½ inches. If you'll add the lengths of the head and legs, you'll find that the sum would exceed the body's length by 6 inches. How tall is this man?

2. I loaned Philip some money, and after he had spent ⅔ of it, he earned ⅔ as much as he had spent, and then had only $20.00 less than what he had originally.

What did Philip owe me?

3. If 6 felines can devour 6 rodents in 1/10 of an hour, how many would it take to devour 100 rodents in 6,000 seconds?

4. A whale's head is just 72 inches long, and his tail is as long as his head and ½ the body, which is ½ of his whole length.

How big is this whale?

5. When Philip's age equals his father's, he will be 5 times as old as his son is now. By then, however, the son will be 8 years older than Philip is now. The combined ages of Philip's father and Philip total 100 years.

What is the age of Philip's son now?

6. An old-time movie house charged admission prices of 25 cents for an adult and 10 cents for a child. If the cashier in the box-office after closing time counted the ticket stubs and found that they totaled 385 while the money amounted to $62.65 — how many children had entered?

7. A farmer sold you ½ an egg more than ½ the eggs that he carried in a basket. He then sold Mr. Haber ½ an egg more than ½ the remainder, and then ½ an egg more than ½ of what was left. Thus 1 egg was left in the basket.

How many did the basket contain originally?

8. A Broadway gigolo very confident of his gambling luck, enters the Star Casino with a certain sum of money. On his first night out he pays $1.50 admission, bets $1.50 and wins $7.00 more than he originally had. The second night out he enters the casino with the money he now had from the previous night, pays $1.50 admission, bets $5.00 and wins $15.00 more than what he had originally when entering the casino on the first night. The third night he enters with the sum of money he then had from the previous night, pays $1.50 admission, and decides to shoot the works, betting all he had left. To his dismay, he lost everything.

How much money did he have originally?

9. One day Phil found to his dismay that he was short in funds by a certain amount. So he wired his sister Amy, as follows:

$$\begin{array}{r} \$ \quad W \; I. \; R \; E \\ M \; O. \; R \; E \\ \hline \$M \; O \; N. \; E \; Y \end{array}$$

How much should Amy send Phil?

10. The life span of a whale is 4 times that of a stork, who lives 85 years longer than a guinea pig, rabbit or a sheep, which live 6 years less than an ox, who lives 9 years less than a horse, who lives 12 years longer than a fowl, who lives 282 years less than an elephant, who lives 283 years longer than a dog, who lives 2 years longer than a cat, who lives 135 years less than a carp, who lives twice as long as a camel, who lives 1,066 years short of the total of all the animals' life spans.

What is each animal's life span?

11. I am told that Esther is twice as old as Paul used to be when Esther happened to be as old as Paul is now. Oh yes, Paul is now 18.

How old do you think Esther is?

12. Mr. Nozitawl loved to know at all times what was going on. One fine day he noticed a laborer digging a hole. Walking over to the man as he was standing in the hole he asked, "How deep is that hole going to be, my good man?" "Well, I'll tell you," said the laborer. "I'm about 2 inches short of 2 yards in height and I'll keep right on digging until the depth reaches twice my height, and then I figure that my head will be twice as far below the ground as it is now above it. Get what I mean, Mac?"

See if you can tell "Nozy" what he means.

13. A group of sidewalk superintendents were watching the progress of a digging job. One thought the job was going too slowly, and said that more men ought to be put to work. This led to an argument as to how fast the job really could be done. One man asked: If it takes 8 men 4 days to dig 2,000 cubic feet of dirt out of a hole in which are to be placed the tanks for a new gasoline station, how many men would be needed to do the job in ½ a day?

14. In a remote country village there lived a poor woman who found it necessary to cross a certain bridge every day in order to earn her livelihood in the next village. One day, as she was about to cross, she was approached by a stranger who promised her riches if she followed his directions. "Take this charm," he said, "and you'll find that each time you cross, your money will be doubled. At the end of each day I'll be waiting for you to pay me ½ the money you then possess. But the charm will bring you this luck for only 3 days. Then you must return it to me or ill-luck will befall you."

After 3 days, the stranger had collected 14 times as much money as the woman had had originally, while she came out $7.00 richer.

Could you find out how much the woman had originally and the amount the stranger was able to collect?

15. You have before you two candles, one of which can last you 5 hours, while the other will last 8 hours. If both were lit simultaneously, how soon would one of the candles be 2.2 times the length of the other?

16. Jack is twice as old as you were when Jack was your age. When you are as old as Jack is now, the sum of your ages will be 100. What are the ages of you and Jack?

17. Doctor Haber, our family physician, strongly urged the two of us to take a pill precisely every half hour, out of the 3 that he gave to each of us. If we followed this advice we would soon be rid of our colds.

For how long did the pills last us?

18. Pop was very anxious to see that his boy made good in the study of arithmetic. So one day he made him a proposition: "Son, here are 26 problems which I'm sure you'll solve correctly. For every one you miss out on, you'll owe me a nickel, but for every one you do get right, I'll give you 8 pennies."

When the problems were finished, no one was indebted to the other.

Find the number of problems the boy solved.

19. "These" and "those" and ½ of "these" and "those" amounts to 7, so, how much is involved in "these" and "those?"

20. Every Wednesday night 3 housewives get together and have a social card game. To break the monotony, they agreed that whenever one of them loses a game, that one would have to double the money of the others. When 3 games had been played and each had lost once, they found that they ended up with $24.00 for each.

How much had each started with?

21. I bought 8 books costing $17.00. Some of the books sold at $1.00, some at $2.00 and others at $5.00 apiece. The subjects the books covered were history, accounting and algebra.

How many at each price did I buy?

22. Professor Haber's wife is 6 years younger than her husband. One day after working on a mathematical formula, he remarked to her: "Did you know, June, that we've been happily married 22 years, and since our marriage our combined ages have exactly doubled?"

How old were they when they were married?

23. A had a number of geese totaling 72, which is exactly 300% that of the number of geese B had. Now, if B bought 6/9 of A's, then B would then have 300% as many geese as A. How many geese did B have originally and finally?

24. At a certain resort 8 people decided to rent a motorboat. Now, if there had been 4 more to share the expense, it would have cost each of the people a dollar less.

Find out what each person had to pay as his share of the rental.

25. If 1/4 of 20 is not 5 but 4, then 1/3 of 10 should be what?

26. John Troublemaker had been drinking too much wine with his dinner at the restaurant. He said to the waiter: I'm a grea' mathematishun, but ther'sh shomething wrong here: every time I add thish bill it comes out diff'rent. Said the waiter: if you're such a good mathematician, solve this problem, and I'll tear up the bill entirely. Here is the problem: A certain wine bottle costs $3.00 more than its cork, and the cost of both is equal to 17 times the cost of the cork.

What is the cost of each?

27. A young chap desirous of buying so many pounds of potatoes, enters a store selling them at 9 cents a pound. However, before buying he makes a few mental calculations and finds himself short by 1,215 pennies of the required price. So he quickly departs and finds another store selling potatoes at 6 cents a pound. Now he finds that after paying for the potatoes he'll have 345 pennies left over.

Could you say how many pounds he really wants to buy?

28. If the assumption is that 2 hens can lay 2 eggs in 2 days, what then would the assumption be as regards the number of eggs that 6 hens could lay in 6/7 of a week?

29. Joe has an appointment to meet me at a certain place and time. If Joe can do 4 miles an hour, he'd be late and I'd have to wait for him 10 minutes, but on the other hand if he increased his rate by 25%, then he'd have to wait for me 20 minutes.

Determine the distance Joe has to cover.

30. The width of a river is 760 feet spanned by a bridge. One-fifth of the bridge stands on one side of the river while the other side holds 1/6 of the bridge.

How long is the bridge?

31. When two numbers are added you'll get 776 62/85, and when subtracted from each other, 510 28/85.

Find these numbers.

32. It is said that A can run a mile in 4.12 minutes, while B can run 4.12 miles in an hour.

Which do you think is the faster runner?

33. Phil starts at a certain time driving his car from New York to Chicago doing 100 miles in an hour. Sixty minutes later, Joe leaves in his car enroute from Chicago to New York, doing 75 miles in an hour.

When the two cars meet, which one is nearer to New York?

34. Little Al was very curious as to the number of chickens Farmer Plunkett had, especially when he was told "if he had as many more, and ½ as many more, and an additional 7, he should have altogether about 32."

Could you satisfy the boy's curiosity?

35. Let us assume that you have 3 dogs and together with their 4 puppies they weigh 528 ounces, or if you happened to weigh 4 other dogs and their 3 puppies and their weight was 592 ounces, what would be the weight of each dog and puppy?

36. I've got 30 $100.00 bills in my wallet. If I happened to spend ½ of this amount today and ½ of the remainder each subsequent day, how long would this money last me?

37. Mr. A can complete a piece of work in 1 minute, B in 2, C in 3, D in 4 and E in 5 minutes.

How long would it take to do the job if all worked together?

38. When Roy went down to the polls to cast his vote, one of the clerks had his doubts as to his true age. "How old are you, son?" he asked. "Oh, about 18." "What do you mean by 'about'?" "I'll tell you: I really gave myself a year less than ¼ of my true age. Satisfied?"

Was the clerk "satisfied?"

39. An ambitious woman goes into a large department store with a certain sum of money and spends it as follows:

1/10 for shoes,

4/9 of the remainder for clothes,

1/10 of the remainder for neckwear,

2/9 of the remainder for a hat,

4/7 of the remainder for a pen and pencil set,

2/3 of the remainder for groceries,

49/50 of the remainder for a bracelet.

Leaving the store she spent 5/10 of the money left for carfare home. Later on her way she bought some candy for 4/5 of the money now remaining. Arriving home she found she had exactly 1 cent left.

How much did she start out with?

40. Suppose I told you that ½ the apples you can see could not be seen, and ⅔ of those not seen can be seen, and that you could see 6 dozen dozen apples more than cannot be seen, and ½ of the apples that cannot be seen could be seen, and ¾ of the apples that can be seen couldn't be seen, and that you would then miss out in seeing ½ a dozen dozen more than you see — well, do you think you could figure out the number of apples that you have?

41. The house numbers of 6 adjoining apartment houses on my block run consecutively in even numbers, and they add up to 9,870. The house I live in has the lowest number. See if you can extract it.

42. As an executive you have on your desk 10 books standing upright, their backs facing you. Each of the books is 3⅞ inches thick. Now, if you replaced the second book with one that is 4½ inches thick, the fourth book with one 5 inches thick and the seventh book with one that is 2 inches thick, how many inches would there be between the front cover of the first book and the back cover of the tenth book, as they now stand?

43. Every Friday Phil has a date to meet his girl in Connecticut who has a beautiful estate. So right after work he takes this train which reaches his destination usually at 5 o'clock. There his girl friend picks him up in her car and rushes him up to her house, arriving 60 minutes later. This procedure was usual every Friday until, one particular Friday, Phil took an earlier train and arrived 60 minutes earlier than usual. After fortifying himself with a cup of coffee and then smoking his pipe, he started walking on the road toward his girl's house. After having walked a certain distance he meets his friend in her beautiful Jaguar, the car is turned around, and in a flash they reach home. This time they arrived 20 minutes earlier than customarily.

How long did Phil walk before he was picked up?

44. A group of out-of-towners decided to have a midnight bite at a local diner after having enjoyed themselves at a Broadway playhouse. Of course, it was understood that each was to share equally in paying the bill, which amounted to $6.00. Then, just as they were ready to leave, they discovered that two of their number had quietly slipped away, leaving them holding the bag, so that each person now had to pay an additional quarter as his share of the bill.

Determine the original number of people in the group.

45. If someone walking along the road travels 30 feet the first minute, 36 feet the second minute, 42 feet the third minute, etc., how long does the road stretch, assuming it takes 720 seconds to reach its end?

46. A friend of mine who had applied for the position of bank teller in a certain bank in this city told me that when he was interviewed for the job, he was asked to solve this practical problem in 60 seconds. If it took him longer, he was told, he would not be accepted for this job. Here it is:

Assuming you're put to work as a teller and someone gave you a check for $63 to be cashed only in 6 bills — no dollar bills or coins — how many of each denomination would you give him?

47. The age of this old man is 87½ years, while his sister is 50. How many years ago was it in which the man was 2.25 times as old as his sister?

48. Washington was admitted as a state into the Union 18 years prior to Oklahoma. By 1911, ¼ of the number of years since Washington was admitted was greater than the number of years' admittance of Oklahoma by only a year and 6 months.

In what year was each state admitted?

49. This man had a yen for walking great distances. Today he's headed for a certain distance, in which he figures that had he gone 1/2 mile an hour faster, he would have walked it in 4/5 of the time. On the other hand, if he'd gone 1/2 mile an hour slower, it would have taken him 2 1/2 hours longer.

How long a walk does he intend to take?

50. The difference in the squares of 2 numbers is 16, and 1 number is exactly 3/5 of the other.

Tell me what the numbers are.

51. A large movie house can accommodate 2,000 people. On this particular night, which was a special occasion, the admission prices were changed to $3.00, $4.00 and $5.00 a seat, and as luck would have it, all seats were sold. At the end of the night's performance, the cashier counted her receipts and found they totaled $7,500.

How many paid admission at each price?

52. Would you please divide 4,700 pennies among A, B and C, so that A would get 1,000 pennies more than B, and B, 800 pennies more than C?

53. Tom bought some turkeys and chickens costing him $112.80. If each turkey cost 28 nickels and each chicken 12 nickels, how many of each did he buy if he bought altogether 108?

54. I need 45 coins to pay Jim the amount of $17.55. He wants it in pennies, nickels, dimes, quarters, half-dollars and in silver dollars.

How many of each coin can I give him?

55. Take 1/7 of the result gotten by adding 12 to a certain number, and you'll find that it is twice the result gotten by subtracting 1 from the number.

What is it?

56. One day as a motorcycle cop was watching the street for speeders, he spied one coming down at breakneck speed about 1/4 mile away. If the cycle covered 55 miles in an hour and caught up to the car in 300 seconds, how fast was the car traveling?

57. Jack showed me his collection of 12 textbooks. Five were on arithmetic and 7 were on algebra. If he allowed me a choice to select 1 book of each subject, how many possible choices could I make?

58. I lent someone as much money as he had and then $10.00 was spent. The second time I lent him as much money as he now had and again $10.00 was spent. The third time I lent him as much money as he now had and again, he spent $10.00. This left him broke.

How much did the borrower start with?

59. There are 8 consecutive odd numbers and when they are multiplied by each other you get 34,459,425.

You'll find them — if you try hard enough.

60. After I had spent $\frac{1}{3}$ and $\frac{1}{4}$ of the money I had in my pocket, I was left with $60.

How much did I have originally?

61. Many history books tell of a certain country having been ruled by a certain number of sovereigns having various names. As a matter of fact, 1/6 were of 1 name, 2/9 were of another, 1/12 were of still another and 1/9 were of each of 2 others. Also, 1/18 were of each of 3 others plus 5 others.

How many do you think there were?

62. If it costs you D dollars to buy T typewriters, what should the selling price be of 1 typewriter, if there's a profit made of P percent, as based on cost?

63. A well-known clothing store purchased about 251 suits for men to sell at $11,795. *Square Deal* suits sell at $35.00, *Sharkie* suits sell at $45.00 and *Famous Brand* suits sell at $55.00 apiece.

How many of each price did the store have?

64. A large piece of timber 100 feet long is split into 3 pieces. If twice the upper piece was short 4 feet of being equal to 3 times the lower piece, and twice the lower piece augmented by 3 feet is equal to the middle piece, how long would each piece be?

65. Three friends met at Yonkers Race-track, and decided to settle up the borrowings they had made of each other on the previous day. Thereupon A gives to B and C as much as each of them has. B gives to A and C as much as each of them then has, and C gives to A and B as much as each of them then has. In the end, each has exactly $6.oo.

How much did each have at the start?

66. Bill, Frank and Tom were very anxious to test themselves in preparation for a coming examination to be given by their college next month. A friend of theirs, who incidentally was a professor, gave each of them a set of problems to work out. Bill was able to solve 9 problems per day and finished his set 4 days before Frank, who solved 2 more problems per day than Tom and finished his set 6 days before Tom got his done.

How many problems were in each set?

67. This fellow, whenever he feels the urge to write a letter, always uses 2 sheets of paper, and for every 12 letters written he discards 1 sheet. If he has available 100 sheets, how many letters could he have written?

68. Nine men earned in a certain time $333.60. If one received $5.00, others $3.75, and the rest, $1.35 for a day's work, how long did they work?

69. My house has a window 8 feet long. Now, if you were to measure from the top of the window to the ground, you'd find it was 31/42 of the height of the house, but if you measured from the window's lower edge to the top of the house, you'd then find it to be 3/7 of the height of the house.

How tall is the window?

70. If somebody wished to pay you off with 39 pieces of money totaling $1.51, some in 3-cent pieces and the rest in 5-cent pieces, how many of each must you take?

71. If giving 5 more apples for a quarter lowers the price a nickel per dozen, what would be the price per dozen?

72. I suppose you'd like to know how old each of this man's sons is. Well, according to him, his oldest son is 4 years older than the second, who is 4 years older than the third, who is 4 years older than the youngest, who is ½ the age of the oldest.

Can you figure it out now?

73. I have 3-cent, 10-cent and 25-cent pieces, and, of the second and third kinds together, I have 310 coins. Well, if the dimes were 3-cent pieces and the 3-cent pieces were dimes, the two kinds would total $17.50. As it is, I have $43. How many coins do I really have of each?

74. Seven men all start together to travel the same way around an island 120 miles in circumference and continue to travel until they all come together again. They travel 5, 6¼, 7⅓, 8¼, 9½, 10¼ and 11¼ miles a day respectively.

How many days till they come together?

75. I've found that a certain number reduced by 6 and the remainder multiplied by 6, will give you the same result if you reduced that certain number by 9 and multiplied the remainder by 9.

See if you can find it.

76. Two classrooms can accommodate 105 pupils consisting of 60 boys and 45 girls. If 10% of the boys and 33⅓% of the girls usually stay absent over a period of time, what would be the percentage of absenteeism for the whole?

77. The owners of a steamboat make $30.00 per day each, but if there were fewer owners by 5, each would make $10.00 per day more.

How many owners are there?

78. A grocer bought some chickens for $18.00 and some turkeys for $48.00. There were 19 more turkeys than chickens and they cost 35 cents more each.

How many of each were bought?

79. How long is a rope from the top of a pole 50 feet high to the top of another 20 feet high, the poles being 16 feet apart?

80. Sam owes me $57.00. If he paid me in $2.00 and $5.00 bills, how many of each type bill could I expect to receive?

81. The good ship *Dingledore* struck an unmarked rock as it neared the coast and began to leak. The crew wanted to take to the lifeboats, but the skipper decided to run for shore. He calculated that he was 40 miles from shore, and that the ship was admitting 3¾ tons of water into her hold every 12 minutes. But if she admits 60 tons she will sink. Her pumps throw out 12 tons of water in an hour, and, if she makes for shore at 2¼ miles an hour, how far from shore will she sink?

82. The combined ages of myself and my parents as of July 4, 1955, total 162 years, 8 months. When I graduated from high school 17 years ago, the combined ages of my parents exceeded mine by 76 years, 6 months, and, when I am as old as my father is now, the combined ages of my parents will exceed mine by 124 years, 6 months.

Can you determine our ages?

83. Would you please tell me (you see, I'm a little absent-minded at times) how much money I originally carried in my wallet, if in spending 1/5 and then 1/5 of what remained, I altogether spent $36.00 somewhere?

84. If you were given $100.00 to buy 100 animals, with chickens at 50 cents, turkeys at $3.00 and calves at $10.00 apiece, how many would you buy of each?

85. There are 2 piles of potatoes, and if for every 6 potatoes in 1 pile there are 7 in the other, and there are altogether 2,990, how many would you say there were in each pile?

86. Twenty people consisting of men, women and girls drank 20 cents worth of milk. If each man paid 3 cents, each woman 2 cents and each girl 1/2 cent, how many people were there of each?

87. If you take 20 dollars from the first and put it into the second of 3 purses, the second would then contain 4 times as much as remains in the first. If 60 dollars of what is now in the second is put into the third, the third will contain twice what is in the first and second together. Now, if 40 dollars be removed from the third and put into the first, there will be ½ as much in the first as in the third.

What did each purse hold originally?

88. A tobacconist sold 100 packs of pipe tobacco, which brought him $15.00. He sold Briggs at 15 cents, Edgeworth at 15 cents, Model at 10 cents, Union Leader at 10 cents, Walnut at 30 cents and Rum & Maple at 25 cents apiece.

How many of each were sold?

89. A tailor cut a 50-yard bolt of charcoal grey cloth into 2- and 3-yard lengths.

How many such combinations of each could he cut out?

90. My friend's aunt was born on November 11th of a certain year. The year she was born in was 28 times that of her age at death.

How old was she when the World War I Armistice was signed?

91. A jeweler had 20 wrist watches to sell during the last week before Christmas. He sold them for $1,084.54. He had Walthams at $30.00, Elgins at $65.00, Longines' at $71.50, Hamiltons at $95.00 and Omegas at $120.00 apiece.

How many of each did he sell?

92. In a typing test Pee Aitch scored 83.2%. If, in scoring, $2\frac{1}{2}$ % is deducted for errors on misspelled words and 1% for miscellaneous errors, could you figure out how many of each type of error was made?

93. Little Joe was sent to buy some stamps at the post office. When he approached the window marked "stamps" he said, "Gimme some 2-cent stamps, 10 times as many penny stamps and the change from the buck in 5-cent stamps."

How many did he buy at each price, and how many were bought in total?

94. Brother Bill had some dimes and quarters totaling 325 cents. One day he received 2 dimes in change when he paid a quarter for a 5-cent pencil, and then found that he had 4 more dimes than quarters.

What was his original amount?

95. The Tomb of Diophantus

Here lies a great man. Mathematically his life is hereon inscribed for all to witness:

During his life as a boy and young man he spent 1/6 and 1/12 of his years.

He entered into matrimony when 1/7 of his life had passed, and 5 years later his son was born. Alas! When his son had only reached 1/2 his father's allotted time on this earth, death took him into its arms. To console himself while grieving over his son's death, he explored and then discovered this science of numbers during a period of 4 years, after which, Diophantus passed on to a better world.

What was his life span?

96. A landlord owns a multiple dwelling housing project consisting of 2-room, 3-room, 4-room and 5-room apartments, renting for $60.00, $90.00, $110.00 and $120.00 respectively, per month. Altogether he has 100 apartments for which he receives $10,500 monthly from his tenants.

Find the number of apartments he can rent of each type.

97. Would you be good enough to find me the number of nickels and $1.00 bills in addition to $5.00 in dimes and $10.00 in half-dollars that I should receive for an equal sum of money in quarters?

98. One-ninth of a log was found stuck in mud, 5/6 was above water, and 2 feet of it was in the water.

How long is it?

99. Two silver cups have only 1 cover for both. If the first weighs 12 ounces, and with the cover weighs twice as much as the other without it, and the second with the cover weighs ⅓ more than the first without it, how much does this cover weigh?

100. A certain fish's tail weighs 9 pounds. Its head weighs as much as the tail and ⅓ of the body combined, and the body weighs as much as the tail and the head combined.

What does the fish weigh?

101. Try and divide 647,100 pennies among 3 people, so that as often as the first gets 500 pennies, the second will get 600 and the third, 700 pennies.

102. Add 8 to a certain number, subtract 8 from the sum, multiply the remainder by 8, and dividing the product by 8 you'll get 4.

Find that elusive number.

103. A father gives to his five sons $1,000, which they are to divide according to their ages, so that each elder son shall receive $20.00 more than his next younger brother.

What is the share of the youngest?

104. A's age when he married was to his wife's age as ¾ to ⅔, but after 12 years their ages are as 13 to 12.

How old were they when married?

105. Find three numbers such that ½ the first, ⅓ the second and 1/5 the third shall together be equal to 39; ⅓ the first, ¼ the second and ½ the third equal to 42; or, 1/5 the first, 1/6 the second and ¼ the third equal to 24.

106. Three brothers, A, B and C, bought a farm for $5,000. A could pay for it alone by borrowing ⅓ of B's money and ½ of C's money. B could pay for it by borrowing ½ of A's and ¼ of C's money. C could pay for it by borrowing ¼ of A's and 1/6 of B's money.

How much money has each?

107. A lady bought a piece of meat for $2.16. If the meat had been a penny a pound dearer, she would have received 3 pounds less for the same money.

How many pounds did she buy?

108. On the road from A to B there are 24 miles uphill, 36 miles downhill and 12 miles level. A courier requires 25 hours to go and 28 hours to return, but if, while going from A to B, there had been 12 miles uphill and 24 miles level, he could have gone in 23 hours.

Find his rate per hour uphill, downhill and on the level.

109. In a certain garrison there were 2,744 men in the cavalry, artillery and infantry together. If ½ the infantry plus 245 equals the sum of 3 times the number of men in the artillery and 6 times the number of men in the cavalry, and if 8 times the number in the artillery and cavalry together would give you the number in the infantry, less 98, how many men were there in each outfit?

110. The Impossible Division Problem

Aunt Jenny had 3 greedy nephews — Philip, Sam and George, who eagerly looked forward to the day when she would die and leave her money to them. But Aunt Jenny decided to play them a little trick. She called in her lawyer one day and made out her will as follows:

The total estate amounting to $1,717 is to be shared by the nephews as follows:

Philip is to receive ½, Sam ⅓ and George 1/9, with the proviso that each is to receive an amount in even dollars only, according to his share. Each nephew is to have 24 hours counting from the hour of death, to calculate the exact amount in dollars he is to receive. If in the event any calculated share amounts to dollars and cents, or if no exact amount in dollars is arrived at at the expiration period of 24 hours, the whole sum becomes forfeited and is to be bequeathed to a worthy charity designated under Paragraph 7 of said will.

The very next day, Aunt Jenny passed away and her lawyer called in her three nephews to hear him read the will. At the end of the reading, they started to calculate their shares, but to their consternation found that no matter how they figured their shares they could not make them come out in an even amount in dollars.

At the end of 23 hours and 50 minutes, they were desperate, but in the last 10 minutes all went well. How did they solve their problem?

111. The Unit Problem

Phil Haber, a buyer for the X, Y & Z Corp., was instructed to purchase 240 pieces of a certain commodity for $200.00. After making a diligent search he found a seller, who offered three different grades of the item, each at a certain price. Now, the number of items bought at each grade was equal in units to the unit price of that particular grade. In other words, if Phil bought 25 pieces at one grade, the unit price would have to be 25 cents, or, assuming 100 pieces of another grade were bought, then the unit price would be 100 cents. How many pieces were bought at each of the three grades, and, what was the unit price per grade?

112. The Apple Problem

Three youngsters each had some beautiful apples to sell. The oldest had 10, the next younger had 30, and the youngest had 50. But, here's the rub: How was it possible for each to sell his apples at the same price and yet receive the same amount of money?

113. The Problem of Problems
(Archimedes' Cattle Problem)

See if you can determine from the following story the number of cattle that once grazed on an island off Italy. The cattle were divided into four herds — white, black, dappled and yellow, and the number of bulls was greater than the number of cows.

PART I

1. The number of white bulls is $\frac{1}{2} + \frac{1}{3}$ of the number of black and yellow bulls.

2. The number of black bulls is $\frac{1}{4} + 1/5$ of the number of dappled and yellow bulls.

3. The number of dappled bulls is $1/6 + 1/7$ of the number of white and yellow bulls.

4. The number of white cows is $\frac{1}{3} + \frac{1}{4}$ of the number in the black herd.

5. The number of black cows is $1/5 + 1/6$ of the number in the yellow herd.

6. The number of dappled cows is $\frac{1}{4} + 1/5$ of the number in the dappled herd.

7. The number of yellow cows is $1/6 + 1/7$ of the number in the white herd.

How are you doing at this point? Were you able to determine the number of each kind of bulls and cows? If so, rejoice but momentarily, as you only show promise as a skilled mathematician, but the real test lies ahead:

PART II

8. The number of white and black bulls equals a Square Number.

9. The number of dappled and yellow bulls equals a Triangular Number.

If you've come up with answers to these also, finding the total number of cattle, then, my friend, you should know that you are highly skilled in numbers and can truly call yourself a mathematician of the first order!

Solutions and Answers

1. 5 feet, 11 inches tall.

2. Assume that 3/3 or 1 represents the whole amount of money lent, then,

1 — 2/3   = 1/3, the money left, and,
2/3 × 2/3 = 4/9, the money earned, therefore,
1/3 + 4/9 = 7/9, the total he now had.

The common denominator is in ninths and 9/9 or 1 represents the original money lent, and 9/9 — 7/9 = 2/9, which is the equal of $20, the amount short of the original amount. To determine what the original amount is, invert 2/9, which becomes 9/2. 9/2 × 20 = $90, the original amount. The proof is as follows:

$90 — (2/3 × 90) = $30,
2/3 × 60   = 40.
                = $70, which is less than $90 by $20.

3. Since an hour contains 3,600 seconds or 60 minutes, 1/10 of an hour would be 6 minutes, and 6,000 seconds is equal to 100 minutes.

Felines	Rodents	Minutes
6	6	6
?	100	100
(1) 6	6	6
1	6	36

(2)	1	6	36
	1	100	600
(3)	1	100	600
	6	100	100

Therefore, it will require 6 felines to devour 100 rodents in 100 minutes or 6,000 seconds.

4. Represent the whale's length by 1. The body is $\frac{1}{2}$ of the length. The tail is $6' + (\frac{1}{2} + \frac{1}{2})$. The length of the head is 6 feet.

$1 - \frac{1}{2} = \frac{1}{2}$, the rest of the length, consisting of $(6' + 6') + (\frac{1}{2} + \frac{1}{2})$. If $12' = \frac{1}{4}$, the rest of the length, then invert $\frac{1}{4}$ to $4/1$ and multiply by 12, giving the whale's length as 48 feet.

5. Let P = Philip's age now
    S = his son's age now
    x = the number of years elapsing

$P + x = 5S,$	$P + x - 5S = 0,$
$S + x = P + 8,$	$P - x - S = -8,$
$2P + x = 100.$	$2P + x = 100.$

$2x - 10S = \quad 0$	$2x + 2S = \quad 16$
$-x \qquad = -100$	$x \qquad = 100$
$\overline{x - 10S = -100}$ (1)	$\overline{3x + 2S = 116}$ (2)

$$x - 10S = -100,$$
$$3x + 2S = \quad 116.$$

$$30S = 300$$
$$2S = 116$$
$$\overline{32S = 416}$$
$$\text{therefore } S = \quad 13$$

Hence the son's age now is 13, and his father, Philip, is 35, who in 30 years from now will be 65 years old, his father's current age.

6. Let A = the number of adults

C = the number of children

25A = the money received from adults

10C = the money received from children

$$25A + 10C = 6265, \qquad -10C = -6265$$
$$A + \quad C = 385. \qquad \quad \underline{25C = \quad 9625}$$
$$ \qquad \quad 15C = \quad 3360$$
$$\text{therefore } C = \quad 224.$$

Hence 224 children entered the movie.

$$224 \text{ children @ } 10\cent = \$22.40,$$
$$161 \text{ adults } \quad @ 25\cent = \underline{\quad 40.25}$$
$$\$62.65$$

7. 15.

8. \$10.

9. $\qquad \$\ 97.62$

$\qquad \quad \underline{10.62}$

$\qquad \$108.24$

10. A camel lives 75 years; a carp, 150; a cat, 15; a dog, 17; an elephant, 300; a fowl, 18; a horse, 30; an ox, 21; guinea pigs, rabbits and sheep, 15; a stork, 100; and a whale, 400 years. The total amounts to 1,141 years.

11. 24.

12. The hole will reach a depth of 8½ feet.

13.

Men	Days	Cubic feet
8	4	2,000
?	½	2,000

(1)	8	4	2,000
	32	1	2,000
(2)	32	1	2,000
	64	½	2,000

Therefore, it'll take 64 men to dig 2,000 cubic feet in ½ day.

14. She originally started with $1, while the stranger collected $2 at the end of the first day, $4 the second day and $8 at the end of the third day or $14 total.

15. In 2½ hours.

16. Jack: 44 4/9, your age: 33 1/3.

17. 1 hour.

18. Let p = the number of problems solved
q = the number of problems not solved
8p = the money due for correct answers
5q = the money to be fined for incorrect answers.

$$8p - 5q = 0, \qquad 5q = 0$$
$$p + q = 26. \qquad \underline{8q = 208}$$
$$13q = 208$$

therefore   q = 16 and
p = 10

Hence, 10 problems solved times 8 cents = 80 cents and 16 problems incorrectly solved times 5 cents = 80 cents, the net difference being zero.

19. There are four possible answers:

  "these": 3 2/3, 2 2/3, 1 2/3, 2/3
  "those": 1,   2,   3,   4

20. $12, $21 and $39.

21. There are at least two possible answers:

    2 @ $1 = $ 2.00,   5 @ $1 = $ 5.00,
    5 @ 2 = 10.00,   1 @ 2 =   2.00,
    1 @ 5 =   5.00 or,   2 @ 5 = 10.00
    ──   ───────   ──   ───────
    8         $17.00   8 $17.00

22. The professor was 25 years old and his wife was 19 at the time of their marriage.

23. Originally B had 24, and finally, 72, which was the original number of geese owned by A.

24. $2.00.

25. 2⅔.

26. If the cost of the cork = 1/17 and that of the bottle, 16/17, then the total cost would be 17/17.

The difference between 16/17 and 1/17 or, 15/17 = the difference in cost between the bottle and its cork, which is $3. If 15/17 = $3.00, then by inverting 15/17 to 17/15 and multiplying by $3 or, 17/15 × $3, you'll get $3.40, the total cost. Hence, 1/17 × $3.40 = 20 cents, the cost of the cork, and 16/17 × $3.40 = $3.20, the cost of the bottle.

27. 520.

28.

Hens	Eggs	Days
2	2	2
6	?	6
(1) 2	2	2
1	1	2
(2) 1	1	2
6	6	2
(3) 6	6	2
6	18	6

Therefore, 6 hens could lay 18 eggs in 6 days.

29. 5 miles.

30. If 1/5 and 1/6 represent both sides of the bridge, then the sum of 1/5 + 1/6 = 11/30, and since 30/30 represents the whole length of the bridge, then the difference of 30/30 and 11/30 or, 30/30 − 11/30 = 19/30, which is equal to 760 feet, the width of the river. Now, if 19/30 = 760, then by inverting 19/30 to 30/19 and multiplying by 760, or 30/19 × 760, you'll obtain 1,200 feet, the full length of the bridge. Therefore, 1/5 × 1,200 = 240 feet, 1/6 × 1,200 = 200 feet and the width of the river = 760 feet, making the total length of the bridge, 1,200 feet.

31. Subtract 510 28/85 from 776 62/85 and divide the remainder by 2. This will result in obtaining one of the numbers, or:

776 62/85 − 510 28/85 = 266 34/85 or 2/5.

266 2/5 = 1332/5 divided by 2 or 1332/5 times ½ = 666/5, which equals 133 1/5, one of the numbers. The difference between 776 62/85 and 133 1/5 gives you 643 45/85 or 9/17. Therefore the other number is 643 9/17.

32. A, as he can run .242718+ of a mile in a minute, while B can do .068666+ of a mile in the same time. The difference between the two is .174052+.

33. They're both the same distance.

34. 100% = the actual number of chickens
100% = 'as many more'
50% = ½ 'as many more'
_____
250% = the total number he would have had, less 7.

Therefore, 250% or 2½, or 5/2 would be equal to the difference of 32 and 7, or 25. Since 5/2 = 25, invert 5/2 to 2/5 and multiply by 25, or 2/5 × 25 = 10 chickens, the number he actually has.

35. Each puppy would weigh 3 pounds and each dog 7 pounds, or 48 and 112 ounces, respectively.

36. Theoretically, I'll always have money.

37. 2 17/60 minutes.

38. Yes; because Roy's age is 22 years and 8 months.

39. $100.00.

40. You *can* see 12 dozen or 144 apples and *couldn't* see 36 dozen or 432 apples, therefore, there's a total of 576 apples that you have or 48 dozen.

41. My house number is 1640. The other five numbers are 1642, 1644, 1646, 1648 and 1650.

42. 30⅞ inches.

43. Ordinarily, the girl friend leaves her house at 4 o'clock and reaches the train station at 5 o'clock just in time to meet Phil who is arriving. Therefore it takes 60 minutes to drive to the station and another 60 minutes to drive home, to arrive at 6 o'clock. The total time would then be 120 minutes to and fro. Now, if they arrived home at 5:40, which is 20 minutes earlier than usual, then the difffference between 120 and 20 minutes represents 100 minutes, or the total time it now took to and fro. If you'll divide 100 by 2, you'll get 50 minutes — the time during which Phil was walking and his girl friend was driving. In other words, it took 50 minutes to meet and another 50 minutes to drive home.

44. Originally there were 8 and each was to have paid 75 cents, but since 2 had slipped away, the remaining 6 each had to pay 75 cents + 25 cents or $1.00.

45. 756 feet.

46. 1 $50.00 bill, 1 $5.00 bill and 4 $2.00 bills.

47. 20 years ago.

48. Washington was admitted in 1889, and Oklahoma in 1907.

49. This time he's all set to walk a distance of 15 miles.

50. Plus or minus 3 and plus or minus 5.

51. Solution by Alligation Alternate:

$7,500/2,000 = $3.75, average admission price

		A	B	C	D	E	F	Answer	
	$3	25	125	1	5	498	5	503 × $3 =	$1,509
$3.75	4	75			3	1494		1494 × ·4 =	5,976
	5		75		3		3	3 × 5 =	15
				4	8			2000	= $7,500

'Alligate' or connect a simple (unit price) that is lower than the average price with a simple that is higher than the average price. The differences between the simples and the average price are then placed inversely in Columns A and B and then proportionately reduced to ratios, which are placed in Columns C and D. Since there are here 4 and 8 parts of a whole, form fractions as follows:

$\frac{1}{4}$ × 1992 to 8 $\frac{5}{8}$ × 8 to 249
$\frac{3}{4}$ × 1992 to 8   + $\frac{3}{8}$ × 8 to 249

Multiples of 4 and 8 must be found in the number 2,000. So, if you'll subtract 8 from 2,000 you'll get 1,992, which is the highest possible multiple of 4, while 8 is the lowest. Multiplying by each of the multiples from 1992 to 8, and from 8 to 249, all answers will then become evident. Therefore, the figures for Columns E and F can now be inserted and by adding across, you'll arrive at the answer.

Obviously, this type of problem admits of many answers. The total number of integral (whole) numbers for answers will be at least 749.

52. Take the difference of what A and B get or $1000 - 800 = 200$, then subtract from 4700, leaving a remainder of 4500. Divide this number by 3 and you'll get B's share of 1500 pennies. Therefore, A will get 2500 and C, 700 pennies.

53. Let T = the number of turkeys
C = the number of chickens

140T = the cost of the turkeys in cents
60C = the cost of the chickens in cents

$$
\begin{array}{ll}
T + C = 108, & 140C = 15120 \\
140T + 60C = 11280. & -60C = -11280 \\
\hline
& 80C = 3840 \\
\text{therefore } C = 48 \\
\text{and } T = 60
\end{array}
$$

54. 5 pennies, 6 nickels, 7 dimes, 8 quarters, 9 half-dollars and 10 dollars. Other answers are possible.

55. 2.

56. 52 miles an hour.

57. 5×7 or 35 ways in which to choose.

58. $8.75.

59. 3, 5, 7, 9, 11, 13, 15 and 17.

60. $\frac{1}{3} + \frac{1}{4} = 4/12 + 3/12 = 7/12$, the portion spent.

Since 12/12 represents the whole amount, subtract 7/12 from 12/12 or, $12/12 - 7/12 = 5/12$, which is equal to the money left or $60.00. Invert 5/12 to 12/5, multiply by 60 or, $12/5 \times 60 = \$144.00$, the original amount. Therefore, 12/12 which is equal to $144.00 subtracted by 7/12 which is equal to $84.00, gives a remainder of 5/12 which is equal to $60.00.

61. 36.

62. One typewriter costs D/T dollars.
D/T times $P/100 = DP/100T$, the profit made.
$D/T + DP/100T =$ the selling price.
$D/T + DP/100T = 100D/100T + DP/100T = (100D + DP)/100T$.
Therefore, $(100D + DP)/100T =$ the selling price of one typewriter.

63. This is an indeterminate problem permitting of many answers, of which one is:

100 "Square Deal"	suits @ $35 =	$ 3,500
1 "Sharkie"	suit @ 45 =	45
150 "Famous Brand"	suits @ 55 =	8,250
251 Suits		= $11,795

64. The upper piece is 31 feet, the middle, 47 feet, and the lower, 22 feet long.

65. A: $9.75, B: $5.25 and C: $3.00.

66. 45. Bill solves 9 problems per day for 5 days, Frank solves 5 problems a day for 9 days and Tom solves 3 problems a day for 15 days.

67. For every 24 sheets used, one is wasted, leaving 23 sheets, and 23 times 4, plus 4, gives 96 sheets, or 48 letters that could be written.

68. 12 days.

69. 48 feet.

70. 22 3-cent pieces and 17 nickels. Other answers possible.

71. 20 cents.

72. The oldest is 24 years old, the next is 20, the next one, 16, and the youngest is 12 years old.

73. 100 3-cent pieces, 250 dimes and 60 quarters.

74. 1,440 days.

75. 15.

76. 20%. 10% $\times$ 60 = 6, the number of boys absent, while 33⅓% $\times$ 45 = 15, the number of girls absent. Therefore, out of a total of 105 pupils, 21 are absent, or 21/105, which reduces to 1/5, or 20%, percentage-wise.

77. 20.

78. 45 chickens and 64 turkeys.

79. 34 feet.

80. Subtract $5 from $57 leaving $52, which is a multiple of 2. $52 divided by 2 will give you the maximum number of $2 bills obtainable, which is 26. Now divide 26 by 5 to determine how many more ways there are of paying by $2 bills. Hence 26 divided by 5 shows that there are 5 more ways —

| $2 Bills: | 26, | 21, | 16, | 11, | 6, | 1 |
| $5 Bills: | 1, | 3, | 5, | 7, | 9, | 11. |

The total number of ways possible is therefore 6.

81. 20 miles.

82. As of July 4, 1955, our ages were:

My age:	34 years, 7 months,
My mother's age:	62 years, 6 months,
My father's age:	65 years, 7 months.

83. $100.00.

84. Solution by Alligation Alternate:
$100 divided by 100 animals = average of $1.

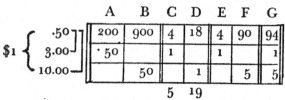

		A	B	C	D	E	F	G
$1 {	.50	200	900	4	18	4	90	94
	3.00	·50		1		1		1
	10.00		50		1		5	5
				5	19			

Connect one of the simples or unit prices that

is lower than the average price with a unit price higher than the average, and insert their respective differences in Columns A and B. The differences shown in Columns A and B are then reduced to their lowest numbers and inserted in Columns C and D. Totaling Columns C and D you will get two ratios, namely, 5 and 19, which are parts of a whole — in this case, of a 100. Two fractions are each then formed from Columns C and D, giving

$$4/5 \times 5 = 4 \atop 1/5 \times 5 = 1 \quad + \quad {18/19 \times 95 = 90 \atop 1/19 \times 95 = 5}$$
$$\overline{5 \text{ Parts}} \qquad \overline{95 \text{ Parts}}$$

95 + 5 equals 100. Therefore, by inserting in Columns E and F the above values, and adding across, you obtain for Column G —

94 chickens	@	.50 = $	47.00
1 turkey	@	3.00 =	3.00
5 calves	@ 10.00 =		50.00
100 animals		= $	100.00

85. There are 1,610 and 1,380 potatoes in each pile.

86. There are 1 man, 5 women and 14 girls.

87. $50.00, $100.00 and $120.00, respectively.

88. 20 packs of Briggs, 15 of Edgeworth, 35 of Model, 10 of Union Leader, 5 of Walnut and 15 of Rum & Maple. Other answers possible.

89. 2-yd. lengths: 22, 19, 16, 13, 10, 7, 4, 1
3-yd. lengths: 2 4 6 8 10 12 14 16.

90. There are two possible answers. The first is directly applicable, while the second is as yet hypothetical.

	(1)	(2)
Date of Birth:	Nov. 11, 1876	Nov. 11, 1904
Age at Armistice:	42	14
Age at Death:	67	68
Year of Death:	1943	1972

91. 10 Walthams, 4 Elgins, 3 Longines', 2 Hamiltons and 1 Omega. Other answers are possible.

92. Inasmuch as he scored 83.2% on the test, 100% − 83.2% = 16.8%, the percentage deducted for the total errors made. There are at least 5 possible answers:

Misspelled words:	2	4	6	3.6	5.6
Miscellaneous errors:	11.8,	6.8,	1.8,	7.8,	2.8
Total words	13.8	10.8	7.8	11.4	8.4

93. 5 2-cent, 50 1-cent and 8 5-cent stamps for a total of 63. Other answers are possible.

94. 10 dimes and 9 quarters.

95. His allotted time was 84 years.

96. There are at least 9 possible answers:

2-Room Apts:	9,	8,	7,	6,	5,	4,	3,	2,	1,
3-Room Apts:	5	10	15	20	25	30	35	40	45
4-Room Apts:	81	72	63	54	45	36	27	18	9
5-Room Apts:	5	10	15	20	25	30	35	40	45.

97. There are 5 nickels, a $1 bill, 50 dimes,

20 half-dollars and 65 quarters. The sum of 65 quarters — $16.25 — is equivalent to the sum of all the other denominations.

98. 36 feet.

99. 3 ounces.

100. Assuming the tail weighing 9 pounds is ⅓ of the body's weight, then 3 times 9 would represent the body's weight of 27 pounds. Then the weight of its head would be represented by the difference of 27 pounds and 9 pounds, giving you 18 pounds as the weight of its head. Therefore, the whole fish weighs 54 pounds.

101. The first will get 179,750 pennies, the second, 215,700 and the third, 251,650 pennies.

102. 4.

103. $160.00.

104. 27 and 24.

105. 30, 48, 40.

106. A: $2,000, B: $3,000 and C: $4,000.

107. 27 pounds.

108. 2 miles uphill, 4 miles downhill and 3 miles level.

109. 98 men in the Artillery, 196 in the Cavalry and 2,450 men in the Infantry.

110. The lawyer, unknown to the nephews, was somewhat of a mathematical wizard. He figured out, that in order for each to receive his share in an exact amount of dollars only, each dividend and denominator must be common to all 3 shares. Thus $\frac{1}{2}$, $\frac{1}{3}$ and $1/9$ have a common denominator of 18, and $1,717 + x$ represents the common multiple or dividend. Therefore,

$$\frac{1,717 + x}{2} + \frac{1,717 + x}{3} + \frac{1,717 + x}{9} = 1,717.$$

Clearing fractions you obtain —

$$15,453 + 9x + 10,302 + 6x + 3,434 + 2x = 30,906$$
$$29,189 + 17x = 30,906$$
$$17x = 1,717$$
$$\text{therefore } x = 101$$

By adding \$101 to \$1,717 you get a common multiple of \$1,818, which is shared as follows:

Philip gets	$\frac{1}{2} \times \$1,818 =$	\$ 909
Sam gets	$\frac{1}{3} \times 1,818 =$	606
George gets	$1/9 \times 1,818 =$	202
Total cash amount of will $=$		\$1,717

10 minutes before the expiration period of 24 hours, the lawyer told the nephews that if they paid him \$101 (for certain technical legal fees), he would gladly give each their due share of Aunt Jenny's estate. Needless to say, since each was at his wits' end trying to figure out his particular share (they had stayed up all night trying to figure out their shares), they readily agreed without any further questions as to

how the lawyer was able to come up with the solution.

111.

Let x = the no. of pieces bought at the 1st grade
x^2 = the price paid
y = the no. of pieces bought at the 2d grade
y^2 = the price paid
z = the no. of pieces bought at the 3d grade
z^2 = the price paid

$$x^2 + y^2 + z^2 = 20,000,$$
$$x + y + z = 240.$$

x + y = 240 − z, or z = 240 − (x + y)
therefore, $x^2 + y^2 + [240 − (x + y)]^2 = 20,000$.

$x^2 + y^2 + 240^2 − 480(x + y) + x^2 + 2xy + y^2 = 20,000$, and adding 2xy to each member, we obtain

$2(x + y)^2 − 480(x + y) = 20,000 + 2xy − 240^2$ then dividing by 2 and completing square, you get $(x + y − 120)^2 = xy − 4,400$.

x + y = 120 and x + y and xy end in a zero; try x + y = 130, 140, etc. to give integral answers. If x + y = 140 then $(x + y − 120)^2 = 400$ and xy − 4,400 = 400 or xy = 4,800.

$x(140 − x) = 4,800$, and $x^2 − 140x + 70^2 = 70^2 − 4,800$. $(x − 70)^2 = 100$, therefore, x = 70 plus or minus 10, y = minus or plus 10 and z = 100. Finally, you arrive at 60 pieces @ 60 cents, which equals \$36, 80 pieces @ 80 cents, which equals \$64, and 100 pieces @ \$1, equaling \$100. Hence you've accounted for 240 pieces bought for \$200.

112. The boys discovered that they should sell their apples in multiples of 7 and charge a nickel for each 7. If an apple was left over it was to be sold for 15 cents. Therefore, the oldest boy sold 7 of his apples for 5 cents and 3 left over for 15 cents apiece. The second sold 28 apples for 20 cents and 2 at 15 cents apiece, while the youngest sold 49 apples for 35 cents and 1 apple for 15 cents. So you see now how each boy earned his 50 cents. Many other answers are possible.

113. This problem is considered one of the most difficult ever to face a mathematician. It is classed as an Indeterminate Quadratic, and one answer reduces to the form of —

$$x^2 - 4{,}729{,}494y^2 = 1.$$

W =	the number of white bulls
B =	the number of black bulls
Y =	the number of yellow bulls
D =	the number of dappled bulls
w =	the number of white cows
b =	the number of black cows
y =	the number of yellow cows
d =	the number of dappled cows

Equations of Part I:

W =	5/6(B + Y),	(1)
B =	9/20(D + Y),	(2)
D =	13/42(W + Y),	(3)
w =	7/12(B + b),	(4)
b =	11/30(Y + y),	(5)
d =	9/20(D + d),	(6)
y =	13/42(W + w),	(7)

The above values resolved are:

B	=	7,460,514
W	=	10,366,482
D	=	7,358,060
Y	=	4,149,387
b	=	4,893,246
w	=	7,206,360
d	=	3,515,820
y	=	5,439,213

Total Number of Bulls: 29,334,443
Total Number of Cows: 21,054,639
Grand Total: 50,389,082

Equations of Part II:

$$W + B = \square, \text{ a Square Number} \qquad (8)$$
$$D + Y = \triangle, \text{ a Triangular Number} \qquad (9)$$

An estimate of the number of cattle was arrived at by mathematician Amthor, who in the year 1860 figured out that 10 to the 206,542d power or 10^{206542} times 766 would give the required number. Also, a surveyor and civil engineer named A. H. Bell tried in the year 1889 to find an answer, with three others. They arrived at 30 of the figures for "y" and 12 for "x."

The total number of figures for "x" turns out to be 206,531. By multiplying each of the numbers in Part I by 4,456,749 you'll arrive at the number of cattle.

Part II has as yet never been fully resolved. Therefore only the following has been computed:

White Bulls:	1,596,510......................341,800
Black Bulls:	1,148,971......................178,600
Dappled Bulls:	1,133,192......................894,000
Yellow Bulls:	639,034......................026,300
White Cows:	1,109,829......................564,000
Black Cows:	735,594......................645,000
Dappled Cows:	541,460......................318,000
Yellow Cows:	837,676......................123,700

Each line of dots as shown above represents 206,532 figures, and the total number of figures in each line is 206,544 or 206,545. For the reader who wants more detailed data, let him refer to the May, 1895 issue of the American Mathematical monthly, where engineer Bell tells you that each of the numbers is *only* ½ mile long! Also, that the complete solution has as yet not been made. He further indicated that 1,000 men working 1,000 years would be required in order to calculate all the possible answers to this problem of problems!